LOOK AT
FUR AND FEATHERS

Franklin Watts Inc.
387 Park Avenue South
New York
N.Y. 10016

US ISBN: 0-531-10720-5
Library of Congress Catalog
Card Number: 89-40008

Design: Edward Kinsey
Illustrations: Simon Roulstone
Phototypeset by Lineage Ltd, Watford
Printed in Italy
by G. Canale & C S.p.A. - Turin

Picture credits:
Animal Photography 19
Bruce Coleman 6
Reg Horlock 4a, 4b, 5a, 5b
Survival Anglia 7, 21
Zefa 8, 9, 10, 11, 12, 13a, 13b, 14, 15, 16, 17,
18, 20, 22, 23, 24, 25, 26, 27, 28

LOOK AT
FUR AND FEATHERS

Henry Pluckrose

FRANKLIN WATTS
London • New York • Sydney • Toronto

When you were born, your body was covered
with a coat of fine, almost invisible hair.

As you grew, your hair grew as well.
Your hair grew thickly on your head.
On the rest of your body, it hardly grew at all.

5

Many animals are born
with very litte hair or fur.

Kitten

Some, like this wildebeest, are born
with a thick coat of fur.

Most birds are naked
when they hatch from the egg.

Some, like ducks and chickens, break from the egg already covered with a feathery coat.

Mammals need hair or fur for warmth and protection.
The thick fur of the polar bear
keeps it warm in the cold Arctic.

The arctic fox is a hunter.
Its white fur helps it to move
almost invisibly across the ice and snow.

The tiger is also a hunter.
The yellow and black stripes on its coat
make it difficult for the animals
it is hunting to spot it coming.
The pattern is a useful camouflage.

Wolves and cheetahs are hunters.
Can you tell from their coloring
where they live and hunt?

Wolves

Cheetah

The color and pattern of an animal's coat
can also give it protection against its enemies.

The dappled coat of the fallow deer looks like
patches of sunlight and shadow found in woodland.

The coat of the giraffe blends
with the trees in the grasslands of Africa.
This is a useful camouflage against attacks.

The color of an animal's coat
helps in other ways.

Badgers have poor eyesight.
Their unusual black and white markings help them
to recognise each other more easily.

The white bob on a hare's tail is used
to warn other hares when an enemy is near.

Some animals are very brightly colored.
The colors on the face of the male mandrill
make it look very attractive to its mate...
and very frightening to an enemy.

Many animals can use their coat of fur to give messages.
What is this cat 'saying'?
What has happened to its fur?

Feathers keep birds warm.
They are also needed for flight.
There are several different kinds of feather.
Wing and tail feathers are used for flying.

Every bird is also covered with a coat
of finer, softer feathers.
These feathers keep the bird warm.

Birds need to keep their feathers clean.
They preen them with their beaks.
They even bathe to remove the small insects,
called mites, which live in their feathers.

The feathers of water birds are very oily.
The oil protects the feathers.
If the feathers became too heavy with water
the bird would drown.

When birds lose their feathers, new ones grow.

Often a bird's coloring changes
as its grows into an adult.
A young swan, called a cygnet,
is not white like its parents.

Peacock

During the breeding season, male birds often use their brightly colored feathers to attract a mate.

Bright colors are easily seen.
The feathers of ground nesting birds
usually blend with their surroundings.
This makes the birds very difficult to see.

Feathers can also help hunting birds.

The barn owl has velvety soft feathers in its wings.

These feathers help it to fly silently as it hunts its prey.

In times past,
animals were often killed for their skins,
and birds were hunted for their beautiful feathers.
The skins and feathers
were used to make clothing.

Can you guess
what might have been used
to make all these things?

Do you know?

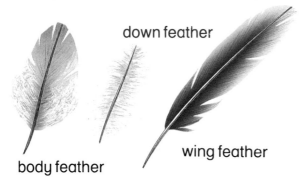

down feather

wing feather

body feather

● There are three types of feather — body feathers, down feathers (which keep the bird warm) and flight feathers.

● The feathers on the center of a bird's tail are symmetrical, but the side feathers on the tail are lopsided, so that they provide maximum lift when air moves across them.

● Flight feathers are narrower on the leading edge. This helps to produce lift as the feathers cut through the air.

● Birds molt (lose their feathers) in a special way. To be able to fly, birds must be perfectly balanced. The loss of feathers on one side of the body is always perfectly balanced by an equal loss on the other side.

Things to do

● Become a bird spotter. Keep a diary and record in it all the different kinds of birds you see in your area. Which birds do you see throughout the year? Which birds are seasonal visitors? Record the different habitats in which the birds live. What kind of birds live in woodland, meadows, in your garden, by a lake or on a cliff top?

Make a quill pen

● A pen cut from a feather is called a quill. The best feathers to use are the strong flight feathers, which come from the leading part of a bird's wing. Turkey feathers are most commonly used, but you could use hen or crow feathers, which you may find on the ground.

Before you cut a quill, the feather must dry out well, but you can practise with new feathers. You will need a sharp penknife and a needle.

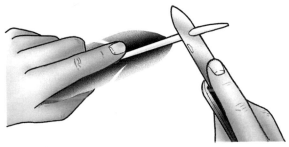

1. Cut a curve into the shaft of the feather about 2cm from its tip.

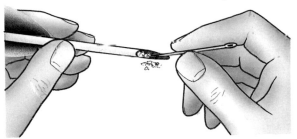

2. Remove the pith from the inside of the shaft with the needle.

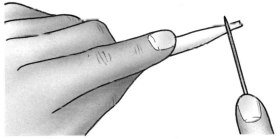

3. Cut a square end.

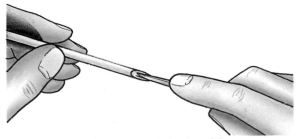

4. Cut a slit near the tip. Now try to write with your quill pen.

Words about fur and feathers

Can you find out what these words mean?

fur trader	feather spray
fur collared	feathergrass
fur below	featherbed
featherweight	feather edged
feathering	feather headed
featherbrain	feather footed
feather duster	feather stitch

Sayings about fur and feather

Can you find out what these sayings mean?

To smooth rumpled feathers
To have a feather in one's cap
To feather one's nest
Birds of a feather flock together
To be in full feather
To stroke the fur the wrong way
To make the fur fly
Oh, my fur and whiskers!

Index